PRO-**LIFE**

PRO-**LIFE**

Saving the Lives of Unborn Children,
Making Possible Their Descendants,
and Helping Their Parents

JIM HARRISON

Xulon Press

Xulon Press
2301 Lucien Way #415
Maitland, FL 32751
407.339.4217
www.xulonpress.com

Printed in the United States of America.
Edited by Xulon Press.

ISBN-13: 9781545604915

Photograph on the front cover: This is a father with his ten-day old daughter. She's substantially the same as she was eleven days before, while still an unborn child.

For each and every unborn child

For the birth parents of unborn children and
adoptive parents

For the volunteers, staff, directors, and supporters of
pro-life pregnancy centers, maternity shelters, and
adoption programs

For pro-life individuals and advocates

And for pro-choice individuals and advocates who,
through informed and compassionate understanding,
may come to value and protect human life from its
very beginning

All of who are among God's wonderful gifts to
the world

TABLE OF CONTENTS

MY WORD TO YOU

Abortion is a subject that can be difficult to write about, and difficult to read about. In putting together this writing, my purpose has been to inform you so that you have a good understanding of important facts, and are better able to form your own opinions for decisions you may have to make. This is what you need to know.

All references and notes for my research have been placed in a section near the back of the book. My research has included some of the finest available medical information including T. W. Sadler's *Langman's Medical Embryology*, Ricki Lewis' *Human Genetics, Concepts and Applications*, Kay Elder and Brian Dale's *In-Vitro Fertilization*, and Elizabeth Crabtree Burton and Richard L. Luciani's *Prenatal Tests and Ultrasound*. I also had the benefit of authoritative work in the pro-life field including Francis J. Beckwith's *Defending Life, A Moral and Legal Case*

Against Abortion Choice, and Randy Alcorn's *Pro-Life Answers to Pro-Choice Arguments* and *Does the Birth Control Pill Cause Abortions?* Some additional, valuable material included Heidi Murkoff and Sharon Mazel's *What to Expect When You're Expecting* and Mara Hvistendahl's *Unnatural Selection: Choosing Boys Over Girls, and the Consequences of a World Full of Men*. I have also included in my research additional, excellent resource material, which is listed in the references and notes section.

It's clear from the title that my own choice would be that of life. However, I value everyone equally, whether pro-life or pro-choice in outlook, and whether one has already counseled or had an abortion, or provided an abortion. (Of course, you could be pro-choice in view, and still choose life for your baby.)

I've tried to be completely honest and accurate in presenting information. I hope you find this worthwhile, and I ask that you read it with an open mind and an open heart.

Jim Harrison

1

THE UNBORN CHILD IS A PERSON

Human life is sacred . . . From its very inception it reveals the creating hand of God.[1]
— POPE JOHN XXIII

I mean the unborn child is a person in the ordinary sense of the word.[2] That is, an unborn child is a human being, and she or he is a human being from the time of conception,[3] the successful result of the process of fertilization at which the nuclei of the female ovum and the male sperm dynamically interact or merge.[4] This is the time pregnancy begins.[5] (Pregnancy begins immediately upon the successful result of the process of fertilization, and before implantation in the uterine wall which takes place approximately one week after fertilization.)

The individual ovum from the mother (which contains twenty-three chromosomes) and sperm from the father (which contains twenty-three chromosomes) cease to exist at this point.[6] Instead, a conceptus is formed, which is a new, although tiny, individual with a human genetic code with its own genomic sequence (with forty-six chromosomes), which is neither the mother's nor the father's.[7] From this point on, through the entire life of this new human being, no more genetic information is needed.[8] All of the inherited characteristics of this special, unique, and unrepeatable person are in place. Our new person's gender, eye color, bone structure, hair color, skin color, susceptibility to certain diseases, and so on have been established.[9] The miracle of life has occurred and begun. It is an established scientific fact that all human life begins at the moment of conception.[10]

Throughout the entire pregnancy, this embryo, later to be called a fetus, will be fulfilling her or his important function of growing within the warmth and comfort of the mother's womb. All that is necessary for this unborn child's growth and development is oxygen, food, water, and healthy interaction with the natural environment.[11] That's not so different from the rest of us. Although the embryo or fetus is in an earlier stage of development, and is living in a special and protected place, we all need oxygen, food, water, and a congenial environment to grow and thrive and continue.

This baby is not just a part of the mother's body; she or he is a human being separate and distinct, but much dependent upon the mother. The unborn child is her or his own person, doing exactly what this child is supposed to be doing at this time of life. The unborn child is not just a potential life; she or he is an actual, real human life with potential to grow in personality and individuality and possess hopes and dreams, and to experience the joy of growth, friendship, and love. The place of residence, for now, is within the mother's womb, and wonderful bonds can develop between the mother and child during the time of pregnancy.

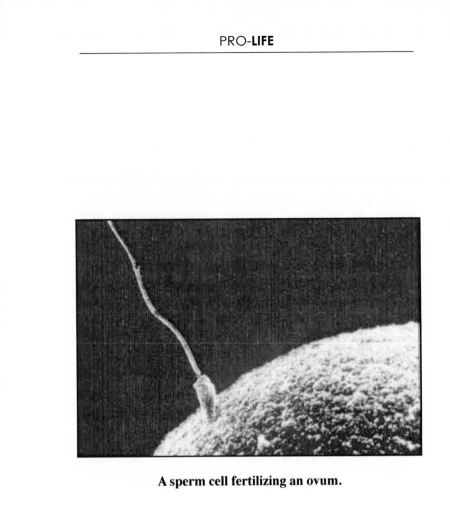

A sperm cell fertilizing an ovum.

2

LIFE IS A MIRACLE

Life is a miracle beyond our compre-
hension, and we should reverence it . . .[1]

—C. J. BRIEJER

It may be hard for us to understand how miraculous
life is because our knowledge, although increasing, is
still quite limited, and because we ourselves are living
in and are part of that miracle. In talking about the
zygote, the one-celled brand new human being formed
upon conception, Bart T. Heffernan, MD, in "The Early
Biography of Everyman," tells us:

The new combination of chromosomes [i.e.,
the zygote's genetic structure] sets in motion
the individual's life, controlled by his own individual code (genes) with its fantastic library

5

of information projected from the past on the helix of ... DNA. A single thread of DNA from a human cell contains information equivalent to six hundred thousand printed pages with five hundred words on a page, or a library of one thousand volumes. The stored knowledge at conception in the new individual's library of instruction is fifty times more than that contained in the *Encyclopedia Britannica*. These unique and individual instructions are operative over the whole of the individual's life and form a continuum of human existence even into succeeding generations.[2]

This look at the beginning human being focuses on the information and instructions contained in the one-celled person. Think about how incredible this is. One cell that is smaller than a grain of sugar,[3] much smaller than a period on this page,[4] contains fifty times more information than that contained in a complete set of *Encyclopedia Britannica*. It links the past, no matter how far back that goes, with the future, no matter how far forward that will go.

The more than 20,000 genes in the human genome are the units of heredity.[5] They are biochemical instructions that tell cells, the basic units of life, how to manufacture certain proteins.[6] Through a variety of mechanisms, a single gene may give rise to many proteins.[7] These proteins control the characteristics that create much of our individuality, from our hair and eye color, to the shapes of our body parts, to our talents, personality traits, and health.[8]

These genes are present in totality in the one-celled zygote.[9] They are all present in every cell, except red blood cells, of the trillions of cells of this human being's body as it develops and goes through life; however, cells differ in appearance and function because they use only some of their genes.[10] Some traits and illnesses are determined by single genes.[11] Most genes do not function alone, but are influenced by the actions of other genes, as well as factors in the environment.[12]

A look at human life from other points of view would also be stunning. Areas of science that are deeply involved in trying to understand human life include molecular and cell biology, biochemistry, embryology, genetics, anatomy, physiology, behavioral sciences, and others. Advanced technologies, engineering, and sophisticated computer and mathematical and statistical techniques support their investigations. Additionally, ethical issues and moral principles have an important role in all we do. **We're not even close to being able to comprehend the complexities of life. Life is special, and we're special, and the developing baby in the mother's womb is special. This miracle is happening around us all the time.**

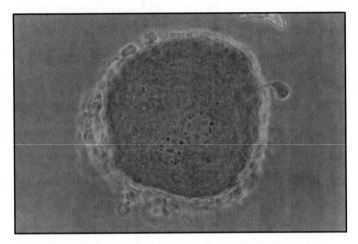

Nuclei of the sperm and ovum dynamically interact to form a zygote.

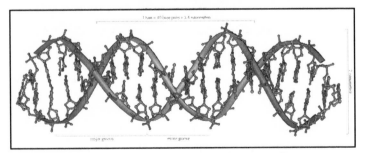

The structure of part of a DNA double helix.

3

DEVELOPMENT OF
YOUR BABY

The human is a magnificent piece of work at all stages of development, wondrous in every regard, from the microscopic until full development.[1]   —STEVE FITZGERALD

It's amazing. The growth and development of the baby, which takes place during the nine months in the mother's womb, is far greater than at any other time of life. What happens, and how it happens, is beyond our most wild imagination. We could never come up with something like this.

The estimated due date for a baby is forty weeks from the first day of the last menstrual period (LMP) based on a twenty-eight day menstrual cycle and, in general, this is considered the length of pregnancy; however, the length of pregnancy is more accurately specified as 266 days, or thirty-eight weeks, after fertilization.[2] (This two week difference is because the expulsion of the egg—ovulation—usually occurs around fourteen days after the first day of the last menstrual period, and it is then that the egg becomes available for fertilization.) For the purposes of the following discussion, age is calculated from the time of fertilization and is expressed in days, weeks, or calendar months.

The baby is called an embryo from the moment of fertilization until the end of the eighth week (the first eight weeks of human development). During the first two weeks, the baby increases in number of cells and travels to and becomes implanted in the womb, which is a protected place to grow. The third to the eighth week is the period during which each of three primary layers of cells begins to grow into specialized parts of the body, including, among others: the digestive system, liver, lungs, heart, sex organs, bones, kidneys, muscles, nervous system, hair, skin, and eyes.[3] As a result of organ formation, major features of body form are established. After that, until birth, the baby is called a fetus. During this time, from the beginning of the ninth week until birth, the tissues and organs of the child mature, and there is rapid growth of the body.

Development of the baby is a highly regulated program of genetic switches that are turned on in specific body parts at specific times.[4] A set of genes controls how the embryo develops its parts in the right places.[5] As the days and weeks of development proceed, different rates of cell division in different parts of the embryo fold the forming tissues into intricate patterns. In a process called embryonic induction, the specialization of one group of cells causes adjacent groups of cells to specialize. Gradually, these changes mold the three primary germ layers into organs and organ systems.[6]

Some things happen so quickly that it sounds remarkable. The future sex cells that will give rise to sperm or eggs for a new generation, that's the generation after this one, begin to group together only seventeen days after this new human life is alive itself.[7] At approximately three weeks after conception, the baby is about the size of an orange seed, and its heart is about the size of a poppy seed, and the heart has already begun to beat, starting at a rate of about seventy times per minute and increasing later.[8] At less than four weeks old, the lungs begin to form.[9] At around six weeks, electrical activity is detectable in the brain, although it has most probably begun before that.[10]

At eight weeks, the baby is well proportioned, and about the size of a thumb. Every organ is present. The liver is making blood, the kidneys function, and the heart beats steadily. The skull, elbows, and knees are forming. The skeleton of the arms and legs and the spine begins to stiffen as bone cells are added.[11] At ten weeks, the baby's fingernails begin to develop.[12] At eleven weeks, the baby can make complex facial expressions and even smile.[13] At twelve weeks, the baby sucks its thumb and practices breathing,[14] since she or he will have to breathe air immediately after birth. At thirteen weeks, facial expressions may even resemble those of the parents. The baby is active, but Mom doesn't feel anything yet.[15] At fifteen weeks, the baby is practicing sucking and swallowing to get ready for breast or bottle feeding.[16] Before four months, the baby will have her or his own unique fingerprints.[17] Before five months, because rapid eye movement (REM) is occurring, the baby is likely dreaming.[18] At five months, Mom may feel her baby kick, turn, or hiccup and may be able to identify a bulge as an elbow or head.[19]

At just over six months, the baby will be able to hear.[20] She or he can listen, learn,[21] and remember. The baby responds differently to her or his mother's voice compared with others, and the baby prefers the mother's voice as it would sound through amniotic fluid. (As a newborn, the baby responds with a calmer heart rate when read a story she or he heard often in the womb.) The child sleeps and wakes, nestling in her or his favorite positions to sleep, and stretches upon waking.[22] A baby's first cry may happen in the womb long before its arrival in the delivery room. Research shows that babies may learn to express their displeasure by crying silently while still in the womb as early as in the twenty-eighth week of pregnancy.[23] At eight months, the pupils of the eye respond to light, and it's getting crowded in there.[24] The baby triggers labor and birth occurs an average of 264 – 270 days after conception. Not until all of these things have occurred on the inside, can we see the new child on the outside.[25]

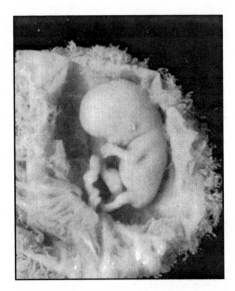

Eight-week-old embryo.

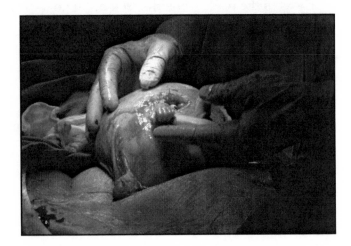

21-week fetus grasping hand of surgeon.

4

THE WAY WE THINK

There is one thing that every individual can do—they can see to it that they feel right. An atmosphere of sympathetic influence encircles every human being; and the man or woman who feels strongly, healthily, and justly, on the great interests of humanity, is a constant benefactor to the human race.[1]

—HARRIET BEECHER STOWE

We've got a problem. It's the way we sometimes think about things.

There are many reasons parents have or have given for considering whether there is room in their lives or on this earth for their unborn child.

They may feel that they are too young, or they're going to school, they were just having fun, have parents who don't understand, are not married, or want to have a life without the complications of a child.

They may have just wanted to see if they could get pregnant, wanted to force their partners to take the relationship more seriously, wanted to force their partners into marriage, perhaps don't want to add stress to the relationship, don't want others to know that they were having sex or got pregnant, or don't want the child in the way of finding their eventual life partners.

They may feel their lives are so demanding already, they're so busy, they're working, they have careers, they don't have the time or energy, they're married with other children who need their care, they need more money, the new child would be a financial burden, they can't see the unborn child, and they don't want to admit the child in the womb is a special human being already with the possibility of a full and wonderful life ahead of her or him.

They may feel they are too old, that the new child is inconvenient, the child may not be born perfect, their significant others don't want the child, the law gives them the right, they can do what they want with their bodies, there are too many people in the world anyway, that there is just no room on earth for this new baby.

We can't help it. It's human nature, to some extent, to focus our decisions on ourselves and what we perceive at the time to be in our best interests. Sometimes we, who are products of our own circumstances, our culture, our times, and our laws, can be unaware of or become desensitized to the plight of others. In this case, we may even become less feeling of our own unborn child, who is so dependent on us for her or his continued existence and opportunity to live its life. In the next chapter, I show some similarities between slavery at an earlier time and abortion today. It all comes down to the way we think; the way of thinking that allowed slavery to exist and continue for so long is similar to how we are sustaining abortion today.

More thought, imagination, and determination would show us that by taking the pregnancy to term and allowing the baby to live, life would be better for us and a full life would be possible for the new child. We have to know and understand the possibilities of keeping the child or using adoptive services, and to get the right kind of support and help when we need it. There are a lot of people and organizations that want to help. **The only bad result is the death of the child, and there are many ways to make sure that doesn't happen.**

5

THE LAW

Forty years ago, seven black-robed men spoke. Hundreds, then thousands, then millions died – all children. Some died from chemical attack, some by dismemberment and the smallest removed from their temporary abode by a vacuum. These who were slaughtered are indeed the world's most vulnerable people. These are the as-yet unborn – children slaughtered by their compliant mothers, the greed of abortionists and the aid of the federal government! A total of 55 million unborn babies have been slaughtered since the seven black-robed men spoke, almost 4,000 each day since that black day in 1973.[1]

—CHARLOTTE BAKER in 2013

Our legal system is extremely important to all of us, and our federal judiciary is fantastic. We citizens of this great country are incredibly fortunate to live under our Constitution and to have in place the federal district courts, appellate courts, and the United States Supreme Court to keep everything in line with our important principles. The judges and justices of our federal courts are among the best of the best this country has to offer. Their intellect, personal standards, and patriotism are of the highest order and serve as models for the rest of us.

Yet they, like the rest of us, are persons of their time and of our culture. And they make mistakes sometimes. Then it takes a while for them and us to fix what we later come to understand are poor decisions.

At the Supreme Court level, consider *Dred Scott v. Sandford* (1857)[2], which dealt with slavery and held that black people have no rights that white people are bound to respect. Or *Plessy v. Ferguson* (1896)[3], which affirmed the constitutionality of legally enforced racial segregation. Or *Korematsu v. United States* (1944)[4], which affirmed the wartime right to exclude American citizens of Japanese ancestry from a west coast area in which their homes were located. Today we look back at these decisions with horror; but then, when they were decided, they made sense to a lot of people, including the justices who decided them.

We can't always count on poor decisions being overturned by the court. Sometimes, even poor decisions don't become explicitly overturned, and *Korematsu* is a good example of that. However, as legal precedent, *Korematsu* is recognized as having limited application.[5] The court can diminish poor decisions without overturning them.

There are strong similarities between how we used to treat blacks in our country, and how we treat our unborn children today. Today, we look at racial prejudices and slavery in particular as absolutely monstrous and inhumane, but not at an earlier time. Think about whether we thought of blacks then, and unborn children now, as property or as individuals who should be protected by the law. In the *Dred Scott* case in 1857, the United States Supreme Court declared all blacks, slaves as well as free, were not and could never become citizens of the United States. In addition, the Court declared provisions of the Missouri Compromise unconstitutional, thus permitting slavery in all of the country's territories. The Court also implied the framers of the Constitution believed blacks "had no rights which the white man was bound to respect; and that the negro might justly and lawfully be reduced to slavery for his benefit. He was bought and sold and treated as an ordinary article of merchandise and traffic, whenever profit could be made by it." It took a long time and a lot of legislation, and Constitutional amendment, to dig our way out of that awful place. It also took changing perceptions and understandings, *Uncle Tom's Cabin*[6] by Harriet Beecher Stowe, and a Civil War to get things right.

Leaving out the war part, we appear to be going through some similar struggles with matters concerning protection of the unborn child. We've got *Roe v. Wade* (1973)[7] which had the immediate effect of nullifying nearly every state abortion law in the country, and additional court decisions and legislation which have followed to interpret and qualify and limit *Roe*. The United States Supreme Court does not consider an unborn child fully developed and just moments before being born, a "person" under the Constitution;[8] even as the courts constructively or fictionally consider a corporation, which is not even a human being, a person within the meaning of the equal protection and due process provisions of the Constitution.[9]

Sometimes the law says what we can get away with. That's the way it was with slavery. That's the way it is with abortion.

We can't always look to the law to help us decide what is right to do or what we should do. We need to look to our consciences, our families, our friends who are mature and have wisdom and care about us, and our places of worship for that.

Slave trade auction block.

Japanese American WWII internment camp.

6

ABORTIONS

Abortion is, in fact, the ruthless killing of an innocent human being. That's what it always has been, and that's what it always will be.[1]
— RANDY ALCORN

Make no mistake about it: abortion is violent. All induced abortion methods involve killing of a human being, and that just doesn't happen without trauma to the body of the child. Below are brief descriptions of some abortion methods. Although it is not my intent to disturb you, the factual descriptions may be troubling. If you understand the first two sentences of this paragraph and if the details of abortion methods make you feel uncomfortable, move on to the next chapter. You already know a lot.

Dismember the Baby

Suction Curettage or Suction-Aspiration. This is the method of abortion that is most commonly used in first trimester abortions.[2] The cervical muscle ring must be paralyzed and stretched open. The abortionist then inserts a hollow plastic tube with a knife-like edge into the uterus. The suction tears the baby's soft body into pieces. The placenta is cut from the uterine wall and everything is sucked into a bottle.[3]

Dilation and Curettage (D&C). This is similar to a suction procedure except a curette, a loop-shaped steel knife is inserted into the uterus. The baby and placenta are cut into pieces and scraped out into a basin. Bleeding is usually heavy with this method.[4]

Dilation and Evacuation (D&E). This type of abortion is done after the third month of pregnancy. The cervix must be dilated before the abortion. Usually Laminaria sticks are inserted into the cervix. These are made of sterilized seaweed that is compressed into thin sticks. When inserted, they absorb moisture and expand, thus enlarging the cervix. A plier-like instrument is inserted through the cervix into the uterus. The abortionist then seizes a leg, arm or other part of the baby and, with a twisting motion, tears it from the body. This continues until only the head remains. Finally the skull is crushed and pulled out. The nurse must then reassemble the body parts to be sure that all of them were removed.[5]

Poison and Burn the Baby

Salt Poisoning (Saline Injection). A long needle is inserted into the mother's abdomen.[6] A concentrated salt solution is injected into the amniotic fluid in which the baby lives. The solution is absorbed by the unborn child through both the lungs and the gastrointestinal tract, poisoning the baby. In addition, the outer layer of skin is burned off by the high concentration of salt. It takes about an hour to kill the baby by this slow method. The mother usually goes into labor about a day later and delivers a dead, shriveled baby, if it is a post sixteen-week saline abortion.[7]

Methotrexate. This is a chemical abortion in which a woman receives an injection of methotrexate, a poison that kills the developing baby. About five days later, she inserts misoprostol tablets into her vagina. The pregnancy usually ends at home within a day or two, although fifteen to twenty percent of women undergoing this procedure require up to four weeks to terminate their pregnancies successfully. The baby and other products of conception that develop during pregnancy are passed out through the vagina.[8]

Starve and Suffocate the Baby

Mifepristone (RU-486). This works by blocking progesterone, a crucial hormone during pregnancy. Without progesterone, the uterine lining does not provide food, fluid, and oxygen to the tiny developing baby. The baby cannot survive.[9] In a few days, the mother uses a second medication, misoprostol. The pregnancy usually ends within four hours after taking the misoprostol. The baby and other products of conception that develop during pregnancy are passed out through the vagina.[10]

Perform a Hysterectomy and Let the Baby Die

Hysterectomy. This is similar to a Cesarean Section, which is usually performed to save the life of the baby. Except, in an abortion, the baby is alive but allowed to die through neglect or sometimes killed by a direct act.[11]

Vacuum the Brain from the Baby

Dilation and Extraction (D&X) or "Partial-Birth Abortion." Using ultrasound, the abortionist grips the baby's legs with forceps. The unborn child is then pulled out through the birth canal and delivered with the exception of its head. While the head is in the womb, the abortionist penetrates the live baby's skull with scissors, opens the scissors to enlarge the hole, and then inserts a catheter. The baby's brain is vacuumed out, resulting in the skull's collapse. The abortionist then completes the womb's evacuation by removing a dead baby.[12] There is now a federal ban on partial-birth abortion, which is enforceable nationwide.[13]

Induce Labor Too Early for the Baby to Survive

Prostaglandin. Prostaglandin is a hormone that induces labor.[14] The abortionist may insert prostaglandin into the vagina or give the medication in the form of an injection to start contractions that will expel the baby.[15] The baby usually dies from the trauma of the delivery. To further ensure that the baby is not born alive, some abortionists use ultrasound to guide them as they inject a drug into the unborn baby's heart to kill the baby. They then administer prostaglandin and a dead baby is delivered. This type of abortion is used in mid- and late-term pregnancies.[16]

Prevent Implantation of the New Baby

The below birth control methods are often referred to as "contraceptives," but they are *not* exclusively contraceptives. That is, they do not always prevent conception. Either sometimes or often they result in the death of already-conceived human beings[17] and in many, and perhaps most cases, this is not understood by the mother and the father of the child.

Intrauterine Device (IUD). The IUD is a small plastic or metal device that is inserted through the vagina and into the cavity of the uterus. In some cases the IUD may prevent fertilization. However, when fertilization does occur, its effect is to prevent the implantation of the tiny new human being into the nutrient lining of the uterus, and effectively kill the baby at one week of life.[18]

Morning-after Pill. This causes a hardening of the lining of the uterus, which prevents implantation of the new human being. In some cases the morning-after pill might act in a sterilizing fashion by preventing an ovulation, and fertilization doesn't occur.[19]

Birth Control Pill. The widely used birth control pills, with their combined estrogen and progestin, have a three-fold mechanism of action, which has been explained in numerous studies and papers: 1) they suppress ovulation [a contraceptive effect]; 2) they thicken the cervical mucus, thereby making it more difficult for sperm to travel to the egg [a contraceptive effect]; and 3) they thin and shrivel the lining of the uterus to the point that it is unable or less able to facilitate the implantation of the newly-fertilized egg [an abortifacient mechanism of action].[20] It's difficult to determine how many abortions are caused by use of birth control pills; however, a sensible estimate based on breakthrough ovulation rates is one abortion every other year for all women on the pill.[21]

Mini-Pill. Progestin-only pills, which have no estrogen, are often called "mini-pills." These pills have as a primary effect making the uterine lining, the endometrium, hostile to implantation by a fertilized ovum. In other words, they cause an abortion of a human being roughly a week after her or his conception.[22] In some cases, they may suppress ovulation.[23]

Injection (Depo-Provera). This is a progestin (medroxyprogesterone) injected every three months. It sometimes suppresses ovulation, but also thins the lining of the uterus, preventing implantation and causing an abortion of the new human being.[24]

Vaginal Ring. The vaginal ring is a soft, flexible, clear plastic ring that is inserted into a woman's vagina, where it slowly releases the hormones estrogen and a progestin for three weeks. These hormones enter into the woman's bloodstream and prevent pregnancy mainly by stopping the ovaries from releasing an egg. It also may thicken the cervical mucus and make the uterine lining thin. The ring's method of action is similar to the combined oral contraceptive pill (the birth control pill). The ring is worn inside the vagina for three weeks, followed by a one-week ring-free interval. At the end of the ring-free week, the woman inserts another ring to begin a new cycle.[25] As in the case of the birth control pill, the third mechanism of action is an abortifacient mechanism.

Skin Patch. This is a patch that sticks to a woman's skin and continuously releases the hormones estrogen and a progestin into the bloodstream. Each patch is worn on the skin for seven days. One patch is worn each week for three weeks. The fourth week is patch-free. Following the seven patch-free days, a new cycle is started when a new patch is applied. The patch prevents pregnancy primarily by stopping the ovaries from releasing an egg, but it may also thicken the cervical mucus (making it harder for sperm to get into the uterus) and make the uterine lining thin.[26] Its method of action is similar to the birth control pill and includes an abortifacient mechanism.

Implant. An implant is a flexible plastic rod about the size of a matchstick that is placed under the skin of the upper arm. It releases a low, steady dose of a progestational hormone to thicken cervical mucus and thin the lining of the uterus. It typically suppresses ovulation as well.[27] It can be used for up to three years and then needs to be replaced.[28] Thinning the lining of the uterus is an abortifacient mechanism.

Side Effects of Reproductive Technologies

Every day of every month of every year, frozen embryos are deliberately destroyed. If several of them are implanted in a woman's womb, some are selectively eliminated. Ironically, the women who participate in this reproductive technology do it because they want at least one baby; yet, several of their other babies may die because of their desire for that one. These threats to the life of the unborn, as those that prevent implantation of the new baby, do not receive as much emphasis because they are not as visible as planned abortions.[29]

[There are true contraceptive birth control methods for men and women and, in addition, natural family planning methods, that are available and do not result in the death of already-conceived human beings. Information on these can be obtained from many sources, including many pro-life pregnancy help centers which are described in Chapter 9 (Help and Support) of this book.]

7

UNBORN BABIES FEEL PAIN

*By acquiescing in an act that can cause such
suffering to a living creature, who among
us is not diminished as a human being?*[1]
— RACHEL CARSON

We've come a long way. In 1973, when Roe v.
Wade was handed down, we knew little about life
in the womb. Today, technology allows us to detect
brain waves as early as six weeks, to see into the
womb through ultrasound and, through fetal surgery,
to correct disabling conditions that a number of years
ago would have condemned a child to an early death.[2]

There is a lot of authority for the understanding
that unborn babies feel pain, and particularly so when
they are the victims of abortion. In open fetal surgery,
in which a pregnant woman's uterus is cut open and

the fetus exposed, surgeons have seen tiny twenty-five-week-old fetuses recoil in what looks like pain when the scalpel is lowered to them.[3] Surgeons have observed twenty-three-week-old fetuses flinching at the touch of the instrument.[4] New evidence has persuaded many doctors that fetuses can feel pain by twenty weeks gestation (that is, halfway through a full-term pregnancy) and possibly earlier.[5] Some research has shown that fetuses as young as eighteen weeks react to an invasive procedure with a spike in stress hormones and a shunting of blood flow toward the brain—a strategy, also seen in infants and adults, to protect a vital organ from threat.[6]

Although the cerebral cortex, which is believed to be the organ of consciousness, is not fully developed in the fetus until late in gestation, a structure called the subplate zone is up and running and some scientists believe this may be capable of processing pain signals.[7] A kind of holding station for developing nerve cells, which eventually melds into the mature brain, the subplate zone becomes operational at about seventeen weeks.[8] There is also some opinion that the brain stem itself can support consciousness and the experience of pain and, like the subplate zone, is active in the fetus far earlier than the cerebral cortex.[9]

The fetus' undeveloped state, in other words, may not preclude it from feeling pain. In fact, its immature physiology may well make it more sensitive to pain, not less, because the body's mechanisms for inhibiting pain and making it more bearable do not become active until after birth.[10]

Then there's the ultrasound videotape and movie of a suction abortion of a twelve-week unborn child. It's entitled *The Silent Scream* and shows the unborn baby dodging the suction instrument time after time, while its heartbeat doubles in rate. When finally caught, its body being dismembered, the baby's mouth clearly opens wide as if screaming.[11]

At ten weeks, if the baby's forehead is touched, she or he may turn its head away.[12] At nine weeks, the entire body is sensitive to touch, except the sides, back, and top of the head.[13] At eight and a half weeks, if the eyelid is stroked, the child squirms.[14] If the palm is stroked, the child's fingers close into a small fist.[15] At seven weeks, the nervous system is well developed.[16] If the area of the lips is stroked, the child responds by bending her or his upper body to one side and making a quick backward motion with a hand.[17]

Even though the unborn cannot yet speak to us with words, they do speak to us through what we see during fetal surgery, our increasing knowledge of the development of structures that may process pain signals, and the babies' reactions and movements upon being aborted or even touched. **We need to understand unborn babies feel pain, and in some cases, they may suffer terribly during abortion procedures.**

8

THE UNENDING IMPACT OF
ABORTION

*Abortion is a horror for the babies who die,
for the women who never fully recover, and
for the people who perform the procedure and
never seek forgiveness; it is a horror for the
lines of descendants eliminated, and for the
society and the culture, which are contami-
nated by their promotion of this social structure of sin.*[1]   —CYNTHIA TOOLIN

Problems After an Abortion

Abortion can change your life in ways you wish
it hadn't. Although the father, grandparents, and
other family members and friends may be truly
affected, the mother of the baby can be hit the

**hardest. For her, there may be immediate compli-
cations from the procedure, longer term medical
issues including her ability to bear healthy children in the future, and a number of near-term
and long-term psychological problems including
depression and post-traumatic stress disorder.**[2]

The most common major physical complications, which can occur at the time of an abortion are
infection, excessive bleeding, embolism, ripping or
perforation of the uterus, anesthesia complications,
convulsions, hemorrhage, cervical injury, and endo-
toxic shock.[3] The most common minor complications
are lesser infections, bleeding, fever, second degree
burns, chronic abdominal pain, vomiting, gastrointestinal disturbances, and Rh sensitization.[4]

While the immediate complications of abortion
are usually treatable, these complications may lead
to long-term reproductive damage of more serious
nature.[5] For example, one possible outcome of abortion-related infections is sterility.[6] Also, women who
acquire post-abortion infections are more likely to
experience ectopic pregnancies later (in which the
fertilized ovum develops outside the uterus, as in a
fallopian tube).[7] Cervical damage from previously
induced abortions increases the risk of miscarriage,

premature birth, and complications of labor during later pregnancies.[8] In addition, premature births, complications of labor, and abnormal development of the placenta, all of which can result from latent abortion morbidity, are leading causes of handicaps such as cerebral palsy and fetal malformation among the newborns post-abortive women may have.[9] Other increased risks to the mother may include death, breast cancer, pelvic inflammatory disease, endometritis (inflammation of the inner lining of the uterus), and cervical, ovarian, and liver cancer.[10]

Women who have multiple abortions face a much greater risk of experiencing physical complications.[11] Teenagers are also at much higher risk of suffering many abortion-related complications, both immediate and long-term reproductive damage.[12]

Abortion has also been linked to increased mental health problems.[13] These may occur soon after an abortion, or after a number of years because many post-abortive women use repression or denial as a coping mechanism.[14] The mental health problems may include depression, nervous disorders, sleep disturbances, regrets about the decision, self-destructive behavior, and feelings of self-hatred, sexual dysfunction, thoughts of suicide and attempted suicide, increased substance abuse (drugs, alcohol, and tobacco), eating disorders (binge eating, bulimia, and anorexia nervosa), child neglect or abuse of children born later, divorce and chronic relationship problems, and symptoms or even clinical diagnosis of post-traumatic stress disorder (in this context, sometimes referred to as post-abortion syndrome).[15]

Women who have had abortions are more likely than others to later become admitted to a psychiatric hospital.[16] Particularly vulnerable to post-traumatic stress and other problems are women who have been coerced or forced into unwanted abortions.[17] As with physical complications, teenagers who have had an abortion are at an especially high risk for mental health problems.[18]

Abortion destroys a woman inside out. It affects her emotionally, physically, psychologically, and spiritually.[19]

There are many roles that the baby's father may have played in the abortion, ranging from opposing the abortion, to not doing enough to prevent the abortion, to not knowing of or wanting the abortion, to being neutral on the abortion decision, to supporting the abortion decision, to abandoning the mother in the face of pregnancy, to coercing the mother to have the abortion.[20] Like the mother, the father may become affected immediately or years later.[21] Some of the feelings and reactions the baby's father may have include sadness, grief, guilt, anger, sense of not being able to protect, holding himself responsible, drug abuse, and alcohol abuse.[22]

It's well understood the most influential person in helping the mother make a decision for life is the father of the baby.[23] As Randy Alcorn, founder of Eternal Perspective Ministries, has stated:

> **Men, we should be the first to accept the responsibility for our actions, the first to stand up for women who need help, and the first to stand up for weak, vulnerable children. When men exercise deep loyalties to women and children, we are at our best; when we violate those loyalties, we are at our worst.[24]**

Loss of Individuals and Descendants[25]

Genealogy has become one of the most popular hobbies in the United States today. People want to construct a family tree, sometimes even a family history, to pass on to future generations. Or they want to find their place in their more extended family, to know who and where they came from, and to find and meet living people whom they had not known were related to them.

Both amateur and professional genealogists develop pedigree charts and/or descendant charts. The pedigree chart starts with one person and works backward through time from the most recent generation (the person's parents) to the last known direct ancestor. A full pedigree chart, that is, one in which all the paternal and maternal direct ancestors are traced into the past as far as possible, comprises sixty-two ancestors within the five most recent generations (two parents, four grandparents, eight great-grandparents, sixteen great-great-grandparents, and thirty-two great-great-great-grandparents). Some genealogists are fortunate enough to be able to trace their direct ancestors through ten or more generations. In ten generations (approximately 300-350 years), you have 2,046 ancestors.

The descendant chart, the second type developed by genealogists, begins with the most distant ancestor known, called a progenitor, and traces all that person's descendants: all children, grandchildren, great-grandchildren through all the generations (as far as possible) to the present day. If the progenitor lived ten generations ago, the descendant chart will likely list an enormous number of people—probably several thousand, at least.

Here is an interesting hypothetical example: an amateur genealogist constructs a pedigree chart to the tenth generation, and finds his most distant direct ancestor (the progenitor), Joshua Patrick Calhoun, who lived in the 1600s. Encouraged by his success in this endeavor so far, he becomes curious about the other people who have descended from Joshua. He decides to do a descendant chart, covering all ten generations of Joshua's descendants, hoping not only to be able to write a family history, but to be able to connect with other people, alive today, to whom he is at least distantly related. Conservatively speaking, if he is successful, he should be able to find several thousand of Joshua's descendants, all of whom share his own DNA in varying proportions.

Instead of staying in the past, let us assume that Joshua Patrick Calhoun is born in September 2010, instead of the 1600s. He might well become the progenitor of several thousand descendants 300 to 350 years from now, that is between the years 2310 and

2360. Think of what all those people could accomplish! Discover a cure for cancer? Start a mining operation on Mars? Build an underwater colony? Think of the great doctors, the great composers, the great scientists, and the brave heroes of the future whose lives would depend upon the birth and life of their progenitor, Joshua Patrick Calhoun.

Now, let us assume, in June 2010, Joshua's mother decided to assert control over her body by aborting her boyfriend's child, who was to have been named Joshua Patrick Calhoun. How many thousands of people will not be born because the abortion machine was turned on at ten a.m. on a bright and sunny day in June? How many great inventions and discoveries will not be made because his mother did not wait until September to deliver him, either to rear him or place him for adoption? Because Joshua Patrick Calhoun was not born, his entire line of descendants will never be.

Many thousands of people will never be born due to this one act of abortion. We all own a slice of forever. What we do has eternal consequences.

Lack of Babies for Adoption

Since the dawn of time, there have been pregnant women who could not parent the child in their wombs, and there have been infertile couples longing for a family. Never has it been harder to bring those two parties together—birth mothers and adoptive parents. **The basic problem is the growing scarcity of babies due to a culture of abortion.** For every eligible baby, an invisible queue of thirty-six couples waits for the chance to take that baby home. There are too few babies to adopt.[26]

It is common for women to believe abortion is a "quick fix" to a big problem. They often believe abortion is cheaper, more confidential, and better for their long-term wellbeing, when in reality, it is quite the opposite, as they don't understand the facts about adoption. Today's adoptions are much different from those in the past and sometimes portrayed in the media. If you, the birth mother, choose adoption, you are in control of the adoption process and get to choose the adoptive family who will raise your child. If you would like, you will receive yearly pictures and letters of your child and may even maintain a relationship with the family and your child. Your adoption services are completely free, and you may also be eligible to receive living expenses from the adoptive family to help you cover pregnancy-related expenses such as rent, food, utilities, transportation, and more. An adoption can also be completed confidentially, where your family, friends, and community won't know about it. **By choosing adoption, you give your baby the gift of a wonderful childhood, adoptive parents who may otherwise have no opportunity or chance at becoming parents, and yourself a better chance of moving on from this difficult time in your life.**[27]

Too Few Women in Areas of the World[28]

For as long as they have counted births, demographers have noted on average, 105 boys are born for every 100 girls. This is our natural sex ratio at birth. The ratio can vary slightly in certain conditions and from one geographic region to the next. More boys are born after wars. More girls are born around the equator, for reasons we don't yet understand. In general, the sex ratio at birth hovers around 105. However, our population is not male-dominated from the start. That more boys are born is itself a form of balance, neatly making up for the fact that males are more likely to die young. Males still account for the majority of soldiers throughout the world. They also disproportionately expose themselves to threats like smoking, a man's pursuit in many countries, or riding motorcycles without wearing a helmet. Boys outnumber girls at birth because men outnumber women in early deaths.

A balanced sex ratio is now considered healthy in most species, to the extent that conservation work often focuses on boosting the number of females. It isn't just that females are the ones who bear offspring, though of course that matters. In mammals that spend years rearing their young, a skewed sex ratio can quickly veer out of control. For example, if females are scarce, males of some species may kill a female's existing offspring to maximize their chance at passing on their genes, inadvertently speeding up the species' path toward extinction. However, when it comes to our own species, we are considerably less attentive. While evolution encourages a balanced sex ratio, our large brains have always worked against one. For as long as we have documented reproduction, we have also sought ways to control it.

In 2007, the booming port city of Lianyungang achieved the dubious distinction of having the most extreme gender ratio for children under five in China: 163 boys for every 100 girls. The numbers may not matter much to the preschool set. However, in twenty years, the skewed sex ratio will pose a colossal challenge. When Lianyungang's children reach adulthood, their generation will have twenty-four million more men than women.

The prognosis for China's neighbors is no less bleak; rampant sex selection abortion has left over 160 million females "missing" from Asia's population. Gender imbalance reaches far beyond South and East Asia, affecting the Caucasus countries, Eastern Europe, and even some groups in the United States (a few thousand is the actual number of fetuses believed to have been aborted on the basis of sex in the United States). As economic development spurs parents in developing countries to have fewer children and brings them access to sex determination technology, couples are making sure at least one of their children is a son. So many parents now select boys that they have skewed the sex ratio at birth of the entire world.

The sex ratio imbalance has already led to a spike in sex trafficking and bride buying across Asia, and it may be linked to a recent rise in crime there as well. More far-reaching problems could be on the horizon: From ancient Rome to the American Wild West, historical excesses of men have yielded periods of violence and instability. The culprit is less deeply rooted cultural gender bias than rising wealth, elite attitudes, and Western influence and technology. Development, at least for the coming decades, will produce not only fewer children overall, but also many fewer girls. The result is a future for many parts of the world, from India to China, Azerbaijan to Albania, where brides are more likely to be bought, women are more likely to be trafficked, and men are more likely to be frustrated. **For the present, we must confront the stark reality that the availability of ultrasound and ready abortion are sharply reducing the number of women in the world.**

The Disproportionate Impact on Minorities

Whatever the intent of the abortion industry may be, by functional standards, abortion is a racist institution. **In the United States, black children are aborted at five times the rate of white children, and Hispanic children don't fare much better.** Abortion is the leading cause of death among black Americans. We can debate the racial intent of Planned Parenthood past and present, but we cannot debate the results. Abortion is by no means an equal opportunity killer.[29] **EW Jackson, an African American, Marine Corps veteran, Harvard Law graduate, and practicing preacher from Chesapeake, Virginia, has affirmed "Planned Parenthood has been far more lethal to black lives than the KKK ever was."[30]**

The Evil of Abortion (The Devil's in the Details)

Every pro-choice argument requires that we pretend, we play mind games, we forget, ignore, or deny the humanity, worth, and dignity of unborn babies. The pro-choice movement thrives on having a silent victim. It thrives on our ability to forget and ignore innocent victims as long as they are out of our sight. Tell a lie often enough and people will eventually believe it and end up reciting it. This is the story of the pro-choice deception in America.[31]

What we have to fear is evil that does not know itself—that is, in fact, convinced it is good. What contaminates one's spirit is the knowledge that this immense engine for evil could not have been constructed and operated in the name of evil. Such energy could only come at the service of some ideal. How could this have happened, one may ask. The answer is, they believed.[32]

The evil that is in the world always comes of ignorance, and good intentions may do as much harm as malevolence if they lack understanding; **the most incorrigible vice being that of an ignorance that fancies it knows everything and therefore claims for itself the right to kill.**[33]

9

HELP AND SUPPORT

Many come to us believing that abortion is their only option, but leave with a fresh confidence ... a new understanding of just how deeply they and their children are valued and loved. They leave with a new resolve to carry their babies to term.[1] —MORNING STAR PREGNANCY SERVICES

Help and support are closer than you think. There are nearly 4,000 pro-life pregnancy help centers in the United States.[2] A typical center employs one or two qualified social workers and has, in addition, ten to fifty or more volunteers.[3] Approximately ninety-eight percent of the staff is female.[4] Some of the services these centers may offer include a twenty-four-hour helpline, pregnancy testing, pregnancy options counseling, ultrasound examinations, parenting support, material aid, referrals for community services including medical and housing, adoption support, mentoring, after abortion support, STD/STI (sexually transmitted disease/sexually transmitted infection) information and testing, and men's programs.[5]

There is a lot of help available through wonderful organizations and people who believe your and your baby's welfare is of the highest importance. It means a lot that the people in these pro-life pregnancy centers who are so concerned about helping so that your baby will be born, are the same people who do the most to help women and men heal after they have had or supported or been affected by an abortion. They are sincere in their belief of the value of your and your baby's lives. You may even benefit from talking to a post-abortive woman who has come to regret her decision and is now serving to help other women make their decision.

This is how easy it is to find a pregnancy center near you, and get contact information. Just call 1-800-712-HELP (4357) anytime, twenty-four hours a day, seven days a week; or go online to www.pregnancycenters.org or www.optionline.

org to find nearby centers by zip code together with addresses, telephone numbers, hours, and services.[6]

So, before you begin down the path of considering an abortion, contact a pregnancy center and discuss your problems and concerns and options with someone who cares. This is a life or death decision for your baby with serious implications for you as well. Also, talk to others you can trust – your partner, your parents, a minister, a priest, a rabbi, an imam or other faith representative, or perhaps a good friend. Become informed; listen to your conscience. Choosing to continue your pregnancy and to parent is challenging. With the support of caring people, parenting classes, and other resources, many women find the help they need to make this choice. You may also decide to place your child for adoption. This loving decision is often made by women who first thought abortion was the only option in their circumstances.

It may even be possible for you to see your baby with an ultrasound examination available at a growing number of the pregnancy centers. This is a simple screening test which uses sound waves of such high frequency they can't be heard by the human ear.[7] There are no known risks and many benefits which have been associated with the use of ultrasound.[8] You may be able to spot your baby's beating heart; the curve of the spine; the face, arms, and legs.[9] You may even catch sight of your baby sucking its thumb.[10] And it may also be possible to determine whether the baby is a girl or a boy.[11]

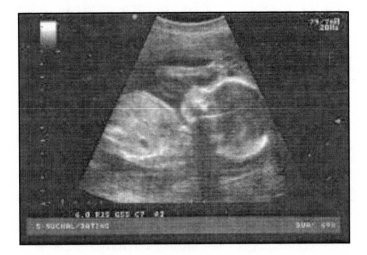

Ultrasound of 22-week fetus.

10

CHOOSING LIFE

Love your children without reserve, for they are a joy and a miracle.[1] —EVAN CLARK

One of the most important things to remember is that your unborn baby is a whole, separate, unique living human being from the time of conception.[2] Your unborn child is her or his own person growing within the warmth and comfort of the mother's womb and doing exactly what this child is supposed to be doing at this time of life. Life truly is a miracle much more special and complex than we can comprehend at this time. The amount of growth and development that takes place during the nine months in the mother's womb, and how it all happens, is an amazing process that is something we could never imagine or come up with on our own. All human life should be greatly valued in all its stages, not for what we do, but for what we are.[3]

There are many reasons parents have for considering whether there is room in their lives for their unborn child. However, more thought, imagination, and determination would show us that by taking the pregnancy to term and allowing the baby to have its opportunity to live, life would be better for us and a full life would be possible for the new child. We have to know and understand the possibilities of keeping the child or using adoptive services, and to get the right kind of support and help when we need it. The only bad result is the death of the child, and there are many ways to make sure that doesn't happen. Even though the law may allow us to seek an abortion under some circumstances, we can't always look to the law to help us decide what is right to do or what we should do. We need to look to our consciences, our families, and our friends who are mature and have wisdom and care about us, and our places of worship for that.

We should keep in mind how violent abortion truly is. All induced abortion methods involve killing of a human being, and that just doesn't happen without trauma to the body of the child. We need to understand unborn babies feel pain, and in some cases, they may suffer terribly during abortion procedures. An abortion can change your life in ways you wish it hadn't. Although the father, grandparents, and other family members and friends may be truly affected, the mother of the baby can be hit the hardest. For her, there may be immediate complications from the procedure, longer-term medical issues including her ability to bear healthy children in the future, and a number of near-term and long-term psychological problems including depression and post-traumatic stress disorder.

It's well understood the most influential person in helping the mother make a decision for life is the father of the baby. As Randy Alcorn, founder of Eternal Perspective Ministries, has stated:

Men, we should be the first to accept the responsibility for our actions, the first to stand up for women who need help, and the first to stand up for weak, vulnerable children. When men exercise deep loyalties to women and children, we are at our best; when we violate those loyalties, we are at our worst.

There are many pro-life pregnancy help centers in the United States. They are staffed by people who believe your and your baby's welfare is of the highest importance. Pregnancy centers near you can be found by calling 1-800-712-HELP (4357) or going online to www.pregnancycenters.org or www.optionline. org. So, before you begin down the path of considering an abortion, contact a pregnancy center and discuss your problems and concerns and options with someone who cares. This is a life or death decision for your baby with serious implications for you as well. You may even be able to see your baby with an ultrasound examination available at a growing number of the pregnancy centers, and benefit from talking to a post-abortive woman who has come to regret her decision and is now serving to help other women make their decision. Also, talk to others you can trust – your partner, your parents, a minister, a priest, a rabbi, an imam or other faith representative, or perhaps a good friend. Become informed; and listen to your conscience.

Choosing to continue your pregnancy and to parent is challenging. However, with the support of caring people, parenting classes, and other resources, many women find the help they need

to make this choice. You may also decide to place your child for adoption. This loving decision is often made by women who first thought abortion was the only option in their circumstances.[4]

I hope you choose life.

Appendix A
ULTRASOUND IMAGING[1]

Ultrasound imaging has become a well-known part of being pregnant and is arguably one of the most powerful tools in prenatal testing practice. It's good news that the vast majority of pregnancies are relatively uneventful and the end result is a beautiful, healthy newborn. However, occasionally some pregnancies are not routine or are considered high risk for a multitude of reasons. Ultrasound images, also known as sonograms, can help to verify the wellness of the baby and, in some cases, help to identify some issues that need attention.

During an abdominal ultrasound, ultra high frequency sound waves, inaudible to the human ear, are transmitted through the abdomen via a device called a probe or transducer to look at the inside of the abdomen. With prenatal ultrasound, the echoes are recorded and transformed into photographic images or video of the baby. The abdominal ultrasound procedures are performed on the surface of the skin using a gel as a conductive medium to aid in the image quality. Studies have shown that ultrasound imaging is not hazardous. There are no harmful side effects to the mother or baby; ultrasound does not use radiation as X-ray tests do. However, there is a general consensus that ultrasound imaging should be reserved for medical purposes, and not done for other reasons in a nonmedical setting. Since ultrasound imaging does transmit energy, it should be treated with respect.

Ultrasound images can be taken in two, three, or four dimensions. Two-dimensional ultrasound images represent a single slice or plane through the body, and the resultant images are in black (fluid including amniotic fluid, blood, urine), white (bone/air), and shades of gray (soft tissue including fat and muscle). Three-dimensional images are boxes or sets of 2-D images that can be rotated and evaluated in any plane. Using specialized software, the 2-D black and white data are manipulated to create the beautiful color 3-D images where you can see the baby's skin and features. Three-dimensional and 4-D ultrasounds are similar; however, 3-D is a still image, and 4-D is a succession of 3-D images over time and shows movement. Two-dimensional is still the standard for diagnostic purposes since 2-D images can provide all the information needed to evaluate a pregnancy properly; however, the ability of 3-D technology to allow views of anatomy in multiple planes has much to offer obstetrics and will likely drive many changes and advancements in the field in the future. A 4-D ultrasound scan makes it possible to observe fetal expressions and behavior, from grimaces to smiles, sucking and licking movements, and limb motion, such as all babies display after birth.

There are a number of times during the pregnancy when an ultrasound examination may be done. The most extensive of these examinations is called an anatomy scan and is usually performed between eighteen and twenty-two weeks of the pregnancy. This measurement of the length of the pregnancy, as are others when we are talking about ultrasound scans, is the number of weeks since the first day of the last menstrual period (LMP), and is usually two weeks greater than the actual age of the baby. References to trimesters are often used and the pregnancy is divided into three trimesters. The first trimester is approximately from weeks zero to twelve, the second trimester is approximately weeks thirteen to twenty-eight, and the third trimester is approximately weeks twenty-nine to forty.

The anatomy scan is extremely important in a pregnancy. Studies consistently show that infant mortality rates decrease in settings where anatomy scans have been performed. Maternal complications are also decreased, especially in cases where the placenta is improperly implanted. It's no surprise that if a problem or serious defect is found in the fetus, the baby has a better chance for survival and successful correction if it is detected before birth. Prenatal detection allows parents and providers to plan for the birth to take place in a hospital that is equipped to care for a sick newborn, and allows the medical team to prepare a strategy for special needs in advance. For instance, if a baby has a heart defect, delivering it in a specialist hospital staffed with pediatric cardiologists ready to move in with the proper equipment in the delivery room, rather than delivery in a small clinic, can mean the difference between life and death.

Most major defects that are found by the anatomy scan will have to wait until birth to be treated. However, there are some temporary interventions that can be performed in utero (while the baby is still in the womb) that can dramatically affect the overall outcome. Doctors can operate on the fetus through a small hole in the uterus using a thin telescopic surgical tool known as a fetoscope. For example, if the fetal kidneys or bladder are blocked, shunts or tubes can be placed to allow the trapped fluid to be drained into the amniotic sac. This can prevent the kidneys or bladder from being over-distended and destroyed. Most of these measures are temporary fixes and surgery will still be required after birth, but they can definitely improve the prognosis.

A good example of surgery before birth for a condition found by the anatomy scan which makes unnecessary the need for later surgery is fetal surgery for an open spinal defect (spina bifida), and surgery for this condition was being performed in the photograph in Chapter 3 (Development of Your Baby) of this book, showing a twenty-one week fetus grasping the hand of the surgeon. A landmark study, co-led by experts at Vanderbilt University Medical Center, proves that babies who have surgery to repair spina bifida while still in the womb have better outcomes than babies who have surgery after birth. The positive outcomes include a decreased risk of death or need for shunt placement in the brain by the age of twelve months to address build-up of fluid on the brain, plus improved mental and motor function.

The anatomy ultrasound scan almost always takes much longer than any other scan in the pregnancy, usually lasting thirty to forty-five minutes, occasionally more. Since there is so much that needs to be evaluated and documented on this scan, examination time is dependent on fetal position and cooperation. The anatomy scan is also time-consuming because the entire fetal anatomy is evaluated and documented. By looking at images of the head and brain, multiple problems including hydrocephalus, or "water-on-the-brain," can be ruled out. In the case of the spine, open spinal defects (spinal bifida) can be ruled out. Observations of the fetal bladder, stomach, and kidneys are made to look for blockages or obstructions. We can be sure that organs are located in their proper places, and that the diaphragm separates the heart from the stomach. Umbilical cord insertion is looked at to be sure the fetal abdomen is closed around the cord insertion site. There should be three vessels identified in the umbilical cord: two arteries and one vein. The fetal heart can be examined to be sure the four chambers are normal, and that the heart is in the proper axis within the fetal chest, and the outflow tracts of the main vessels in the heart are evaluated to ensure that the vessels arise from the expected places. The fetal face is examined, with special attention to lips where a cleft lip can be visualized or ruled out. In the extremities, the bones are counted and measurements made of the long bones to be sure they are the appropriate size. The feet are examined to rule out malformed feet or a clubfoot.

In addition to focusing on the anatomy of the baby in an anatomy scan, a general look will be taken at the amniotic fluid volume and the uterus. They will also look at the mother's cervix, searching for signs of shortening or dilating, which can signal a preterm birth. The location of the placenta is also evaluated and documented. The placenta can be located anywhere in the uterus; a problem only arises if it is found to be blocking the cervix where the baby needs to eventually exit. If any part of the placenta is seen to be close to or covering the cervix, it is called a low-lying placenta or a placenta previa, and it will need to be monitored closely with follow-up ultrasounds. If it were still low after thirty-six weeks, a cesarean section would need to be performed. Low-lying placentas are not uncommon in the second trimester, but only one in 200 pregnancies have a true placenta previa at the time of birth. If a mother is carrying twins or multiple gestations, the entire fetal survey will be repeated for each fetus. It is important not to underestimate the medical value of this examination. The anatomy ultrasound scan is also notable for being the time when the sex of the baby can be reliably revealed.

There are other times during the pregnancy when an ultrasound scan may be performed. The main purpose for an early first trimester scan is to prove that there is an actual live embryo developing, and that it has a heartbeat. The greatest and most reassuring hurdle in any pregnancy is the development of the heartbeat, and ultrasound in the early first trimester can detect a fetal heartbeat as early as six weeks from the first day of the last menstrual period. Additional benefits are to check to see the baby is in the right position inside the uterus, establish an accurate due date, determine the reason for any spotting or bleeding the mother may have, and show how many babies are being carried. Scans in the first trimester can sometimes be done abdominally (when the probe is placed on the skin below the navel), but most of the time, better images are obtained using a vaginal approach. This is called a transvaginal ultrasound and here the transducer (or "probe" or "wand") is shaped like a tampon and is covered with a condom-like protective cover. Gel or lubricant is applied to the end of the covered probe, and the probe is then inserted into the vagina with minimal discomfort. The transvaginal probe has a much higher resolution than the abdominal scanner; therefore it is able to visualize smaller structures with much greater detail. The probe only enters the vaginal canal, does not disturb the cervix or the uterus, and is perfectly safe, even if the mother is experiencing some bleeding or cramping.

Another ultrasound scan that is performed as part of first trimester screening at approximately twelve weeks is a nuchal translucency scan, whose main purpose is obtaining a nuchal translucency (NT) measurement. This twelve-week scan is part of a relatively recent development in first trimester screening, aimed at finding those fetuses that are at higher risk for being chromosomally abnormal. The NT is a small collection of fluid that lies just under the skin at the fetal neck. It is considered thickened if it is greater than three mm, and this indicates the fetus is at higher risk for a chromosomal abnormality. If the fetus is found to be chromosomally normal, but has a thickened NT, careful monitoring will be necessary, because these fetuses are at higher risk for anatomical abnormalities or syndromes. However, in the majority of cases, the fetus is still likely to be normal. Most of the time, the NT scan is done abdominally. Occasionally, for various reasons such as maternal size or fetal position, a transvaginal ultrasound might be performed. At the nuchal translucency scan, in addition to the NT measurement, the fetus will be measured from crown to rump, the heart rate will be documented, and basic anatomical structures such as the hemispheres in the brain, the limbs, abdominal wall, stomach, and bladder will be observed. The scan will also verify the amniotic fluid volume, placenta, uterus, and maternal ovaries look normal. In order to obtain final results for this first trimester screening, the NT measurement taken by the sonographer on ultrasound is combined with lab results on a blood sample taken from the mother.

A third trimester scan is not always recommended in every pregnancy. Once the anatomy scan at approximately twenty weeks of the pregnancy has demonstrated a normal-looking fetus with all the appropriate body parts, there is usually no need to reassess the anatomy as it is not likely to change. However, in the general routine testing of the mother and for various other reasons, sometimes the health care provider will want to take another look with the ultrasound. The primary medical reasons to have the third trimester ultrasound are to check the growth of the fetus, assess the amniotic fluid volume, check the position (breech versus head-first), or to look for signs of fetal distress. By thirty-six weeks, the fetus is expected to have assumed the head-down position (also known as vertex or cephalic presentation) where it will remain until birth. If the provider suspects the fetus is not head-down, or is unsure as to where the head is, they might order an ultrasound to confirm the position. Toward the end of the pregnancy, if the head is not vertex, ultrasound can be used to guide the doctors, should they decide to manually turn the baby around in preparation for a vaginal birth. If the location of the placenta has been a concern earlier on, the provider will order an ultrasound to see whether the cervix is clear. If special attention needs to be paid to the cervix, the scan might be performed transvaginally, but otherwise the evaluation of the fetus in the third trimester will be done transabdominally. Late in the pregnancy, a targeted ultrasound can be performed to assess fetal wellbeing, known as a biophysical profile. This is most likely to be done if the pregnancy goes beyond the due date and helps the sonographer or doctor calculate an amniotic fluid index and look for fetal movements and sustained fetal "breathing." The fetus is not breathing air yet, but if you watch the chest area, you can see the diaphragm moving up and down, and the abdomen and ribs contracting and expanding. Seeing these movements is reassuring because it indicates that the fetus is receiving adequate oxygen at the deepest parts of the brain, and is able to expend energy moving, exercising, and building the muscle tone it will need to breathe and function well outside of the womb. In addition, if an early birth (before thirty-seven weeks) is likely to be needed, doctors can use ultrasound to guide an amniocentesis, or sampling of amniotic fluid, which can be chemically analyzed to definitively determine if the lungs are mature.

Appendix B

ABORTION AND THE DEBATE ON "WHEN HUMAN LIFE BEGINS"

There are some differences of opinion for the point at which human life begins and when it should be protected. For some, the beginning of human life coincides with the formation of a diploid body in which the male and female chromosomes are brought together[1] (the time of conception). For others, true human life occurs after implantation of the embryo in the uterine mucosa.[2] Others believe a new individual is formed only after differentiation of the neural tube[3] (in an embryo, a hollow structure from which the brain and spinal cord form). Others have believed human life begins upon quickening, an old term for when the mother first becomes aware of the movements of the child within her,[4] while others believe life begins when a fetus can live outside the uterus[5] (viability). Still others believe human life begins when a child is born and takes her or his first breath[6] and, in its most extreme form, some individuals consider the acquisition of self-awareness of the newborn to define a new life.[7]

Human life should be protected in all of its stages. The truth is, there is no sensible debate on when human life begins; science has answered that question. Human life begins at conception.[8] That is the only objective point of origin for any human being—the only point at which there was not a human being a moment ago, and there is now.[9]

What science tells us about the unborn, much of it made visible by modern ultrasound imaging technology, provides a real, accurate, truthful window to understanding the full humanity of the unborn.

Education with honesty, and real understanding, are our greatest hopes to stop the dismembering, crushing, scraping, starving, suffocating, burning, poisoning, piercing, and vacuuming (see descriptions of abortion methods in Chapter 6) that is taking place routinely in the abortion clinics and facilitated in some doctors' offices which are the killing fields of today's United States, under cover of some of our laws. That is a primary goal of pro-life advocates and the purpose of this book.

Appendix C

PRAYER FOR ABORTED CHILDREN[1]

Heavenly Father, I pray each child whose precious life has been ended while in the womb of their mother and in violation of your divine plan, is with you in heaven.

I pray they know they would have been born wonderful and beautiful children. I ask you take these children into your arms and comfort them. I pray for forgiveness for their parents and others who counseled or provided or did not do enough to prevent their abortion. I pray these children continue to be with each other and with you.

Each of these girls and boys was fulfilling their important function of growing within the warmth and comfort of their mother's womb. Each one would have been born with their own personality and individuality and hopes and dreams. None of them will ever experience the joy of growth, friendship, achievement, and love on earth. None of them will have children or grandchildren and for that, we are doubly and triply impoverished by their loss.

I pray these girls and boys know how desirable, beloved, and dear they would have been for us, had they had the chance to touch us with their beauty and innocence and become a member of one of our families. I pray in time they become reunited in love with their parents and siblings.

I pray you use the deaths of these children as a catalyst for many graces, by inspiring their parents and others to work for life.

I pray you forgive us all for not sufficiently appreciating and valuing the children of your creation, so that we have allowed the knowing termination of their innocent lives.

I also pray we continue to learn and increase our compassion, so that our choices are always of life.

Through Christ our Lord, Amen.

REFERENCES AND NOTES

Chapter 1. The Unborn Child is a Person

1. Pope John XXIII as cited in *Human Vitae* (On Human Life), Encyclical Letter by Pope Paul VI, July 25, 1968.

2. *Webster's New World College Dictionary*, 4th Edition (Cleveland: Wiley Publishing, Inc., 2006). Definition of "person" as a "human being."

3. Francis J. Beckwith, *Defending Life, A Moral and Legal Case Against Abortion Choice* (New York: Cambridge University Press, 2007), 65.

4. Ibid., 66.

5. Ibid.

6. Ibid.

7. Ibid., 67.

8. Ibid.

9. Ibid.

10. Dr. Hymie Gordon, professor of medical genetics and physician at the Mayo Clinic, as cited in Beckwith, *Defending Life*, 68.

11. Beckwith, *Defending Life*, 67-68.

Chapter 2. Life is a Miracle

1. Dr. C. J. Briejer as cited by Rachel Carson in *Silent Spring*, 40th Anniversary Edition (New York: Houghton Mifflin Company, First Mariner Books Edition, 2002), 275.

2. Bart T. Heffernan, M.D., as cited by Francis J. Beckwith in *Defending Life, A Moral and Legal Case Against Abortion Choice* (New York: Cambridge University Press, 2007), 70.

3. Heritage House '76, Inc., *Milestones of Early Life*, www.abortionfacts.com, viewed 2008.

4. Heidi Murkoff and Sharon Mazel, *What to Expect When You're Expecting*, 4th Edition (New York: Workman Publishing Company, 2008), 121.

5. Ricki Lewis, *Human Genetics, Concepts and Applications*, 8th Edition (New York: McGraw-Hill, 2008), 2.

6. Ibid.

7. T. W. Sadler, *Langman's Medical Embryology*, 10th Edition (Philadelphia: Lippincott Williams & Wilkins, 2006), 5.

8. Lewis, *Human Genetics*, 2.

9. Beckwith, *Defending Life*, 67, 70. Lewis, *Human Genetics*, 50.

10. Lewis, *Human Genetics*, 4.

11. Ibid., 7.

12. Ibid.

Chapter 3. Development of Your Baby

1. Kansas State Senator Steve Fitzgerald as cited in FoxNews.com article *Kansas lawmakers pass measure banning sex-selective abortions*, April 6, 2013, Associated Press.

2. T. W. Sadler, *Langman's Medical Embryology*, 10th Edition (Philadelphia: Lippincott Williams & Wilkins, 2006), 89, 90.

3. Heidi Murkoff and Sharon Mazel, *What to Expect When You're Expecting*, 4th Edition (New York: Workman Publishing Company, 2008), 121.

4. Ricki Lewis, *Human Genetics, Concepts and Applications*, 8th Edition (New York: McGraw-Hill, 2008), 60.

5. Ibid., 53.

6. Ibid., 56.

7. Heritage House '76, Inc., *Milestones of Early Life*, www.abortionfacts.com, viewed 2008.

8. Murkoff and Mazel, *What to Expect*, 150-151. Heritage House '76, Inc., *Milestones of Early Life*.

9. Heritage House '76, Inc., *Milestones of Early Life*.

10. Francis J. Beckwith, *Defending Life, A Moral and Legal Case Against Abortion Choice* (New York: Cambridge University Press, 2007), 71.

11. Heritage House '76, Inc., *Milestones of Early Life*.

12. Ibid.

13. Ibid.

14. Lewis, *Human Genetics*, 57.

15. Heritage House, '76, Inc., *Milestones of Early Life*.

16. Murkoff and Mazel, *What to Expect*, 199.

17. Ibid., 232.

18. Beckwith, *Defending Life*, 72.

19. Heritage House '76, Inc., *Milestones of Early Life*.

20. Ibid.

21. Beckwith, *Defending Life*, 72.

22. Heritage House '76, Inc., *Milestones of Early Life*.

23. Jennifer Warner, *Babies May Start Crying While in the Womb*, WebMD, September 14, 2005.

24. Heritage House '76, Inc., *Milestones of Early Life*.

25. Ibid.

Chapter 4. The Way We Think

1. Harriet Beecher Stowe, *Uncle Tom's Cabin* (New York: Barnes & Noble Books, 2005), 438. [*Uncle Tom's Cabin* was serialized between 1851 and 1852, and published in volume form in 1852.]

Chapter 5. The Law

1. Charlotte Baker in a Letter to the Editor of The Augusta Chronicle titled "Here are some more victims", September 6, 2013.

2. *Dred Scott v. Sandford*, 60 U.S. (19 How.) 393 (1857).

3. *Plessy v. Ferguson*, 163 U.S. 537 (1896).

4. *Korematsu v. United States*, 323 U.S. 214 (1944).

5. Judge Marilyn Hall Patel, U.S. District Court for the Northern District of California, 1984.

6. Harriet Beecher Stowe, *Uncle Tom's Cabin*. [*Uncle Tom's Cabin* was serialized between 1851 and 1852, and published in volume form in 1852.]

7. *Roe v. Wade*, 410 U.S. 113 (1973).

8. Ibid.

9. A corporation is a "person" within the meaning of equal protection and due process provisions of the United States Constitution. Allen v. Pavach, Ind., 335 N.E.2d 219, 221; Borreca v. Fasi, D.C. Hawaii, 369 F.Supp. 906, 911.

Chapter 6. Abortions

1. Randy Alcorn, as quoted in May 2013 *Insight* newsletter of 1st Choice Women's Health Center, Lansdowne, Virginia (a ministry of Life Line, Inc.).

2. Heritage House '76, Inc., *Learn Abortion Facts*, www.abortionfacts.com, viewed 2008.

3. Heritage House '76, Inc., *How Are Abortions Done?*, www.abortionfacts.com, viewed 2008.

4. Ibid.

5. Ibid.

6. Heritage House '76, Inc., *Learn Abortion Facts*.

7. Francis J. Beckwith, *Defending Life, A Moral and Legal Case Against Abortion Choice* (New York: Cambridge University Press, 2007), 87-88.

8. Ibid., 86-87.

9. Heritage House '76, Inc., *How Are Abortions Done?*

10. Beckwith, *Defending Life*, 87.

11. Ibid., 89, 91.

12. Ibid., 88.

13. Partial-Birth Abortion Ban Act (P.L. 108-105). *Gonzales v. Carhart*, 550 U.S. 124 (2007).

14. Heritage House '76, Inc., *How Are Abortions Done?*

15. Beckwith, *Defending Life*, 87.

16. Heritage House '76, Inc., *How Are Abortions Done?*

17. Randy Alcorn, *Does the Birth Control Pill Cause Abortions?*, 8th Edition (Sandy, OR: Eternal Perspective Ministries, 2007), 177.

18. John C. Willke and Barbara H. Willke, *Why Can't We Love Them Both, Questions and*

Answers About Abortion (Cincinnati: Hayes Publishing Company, 1997), 127.

19. Ibid., 128.

20. Alcorn, *Does the Birth Control Pill Cause Abortions?*, 23. Bogomir M. Kuhar, *Infant Homicides Through Contraceptives*, 5th Edition (Bardstown, KY: Eternal Life, 2003), 41-42.

21. Alcorn, *Does the Birth Control Pill Cause Abortions?*, 92, citing J. C. Espinoza, M.D., *Birth Control: Why Are They Lying to Women?* (Human Life International, 1980), 28.

22. Alcorn, *Does the Birth Control Pill Cause Abortions?*, 177-178.

23. Ibid., 177, citing *Drug Facts & Comparisons*, 1996 edition, 419.

24. Ibid., 176.

25. The Society of Obstetricians and Gynecologists of Canada, *Birth Control - Hormonal Methods*, www.sexualityandu.ca/birth-control/hormonal-methods, viewed October 5, 2014.

26. Ibid.

27. Mayo Clinic, *Tests and Procedures - Implanon (contraceptive implant)*, www.mayoclinic.org/tests-procedures/implanon/basics/

definition/prc-20015073, viewed October 11, 2014.

28. WebMD, *Birth Control Implants*, www. webmd.com/sex/birth-control/birth-control-implants-types-safey-side-effects, viewed October 11, 2014.

29. This was taken from the article by Cynthia Toolin, Ph.D., "Assaults On The Unborn: Another Indication of Their Long-Term Impact", Family Resources Center News, Peoria, Illinois, January/February 2012. Previously printed in *Social Justice Review*, March/April 2011.

Chapter 7. Unborn Babies Feel Pain

1. Rachel Carson speaking of animal life suffering and death due to insecticide use, *Silent Spring*, 40th Anniversary Edition (New York: Houghton Mifflin Company, First Mariner Books Edition, 2002), 100.

2. Laura Echevarria, LifeNews.com Editor, *Abortion Backers Can't Ignore New York Times Story That Babies Feel Pain*, February 14, 2008.

3. Annie Murphy Paul, *The First Ache*, <u>The New York Times</u>, February 10, 2008.

4. Ibid.

5. Ibid.

6. Ibid.

7. Ibid.

8. Ibid.

9. Ibid.

10. Ibid.

11. *The Silent Scream* DVD available from Heritage House '76 at www.heritage-house76.com.

12. Francis J. Beckwith, *Defending Life, A Moral and Legal Case Against Abortion Choice* (New York: Cambridge University Press, 2007), 90.

13. Ibid.

14. Ibid.

15. Ibid.

16. Ibid.

17. Ibid.

Chapter 8. The Unending Impact of Abortion

1. Cynthia Toolin, Ph.D., *Assaults On The Unborn: Another Indication of Their*

Long-Term Impact, Family Resources Center News, Peoria, Illinois, January/February 2012. Previously printed in *Social Justice Review*, March/April 2011.

2. Elliot Institute, www.afterabortion.org, viewed 2008.

3. Ibid.

4. Ibid.

5. Ibid.

6. Ibid.

7. Ibid.

8. Ibid.

9. Ibid.

10. Ibid.

11. Ibid.

12. Ibid.

13. Ibid.

14. Ibid.

15. Ibid.

16. Ibid.

17. Ibid.

18. Ibid.

19. Yvonne Florczak-Seeman as cited by Carolee McGrath in article "A Legacy of Heartache" in January 2013 issue of *Columbia Magazine*.

20. National Office of Post-Abortion Reconciliation and Healing, *Description of Aftermath – Impact of Abortion on Men*, www.menandabortion.info, viewed 2008.

21. Ibid.

22. Ibid.

23. Maggie Downing, Executive Director of Mosaic Virginia, in email the subject of which is *Help Us Build our Men's Ministry!*, May 24, 2016.

24. Randy Alcorn, *Pro-Life Answers to Pro-Choice Arguments* (Colorado Springs: Multnomah Books, 2000), 354.

25. This was taken from the article by Cynthia Toolin, Ph.D., *Assaults On The Unborn*. [The calculation of 2,046 ancestors in 10 generations (approximately 300-350 years) was provided by the author of this book.]

26. Paula Rinehart, "Why So Many Families Who Want To Adopt Can't", August 18, 2016, http://thefederalist.com/2016/08/18/why-so-many-families-who-want-to-adopt-cant/, viewed March 14, 2017.

27. American Adoptions, *Abortion or Adoption – Know the Facts Before Making a Decision.*

http://www.americanadoptions.com/pregnant/
deciding_between_abortion_or_adoption,
viewed March 14, 2017.

28. Important sources of this material are *Unnatural Selection: Choosing Boys Over Girls, and the Consequences of a World Full of Men*, by Mara Hvistendahl (New York: PublicAffairs, a member of the Perseus Books Group, 2011), and the review of *Unnatural Selection* by Anne-Marie Slaughter, professor of politics and international affairs, Princeton University.

29. www.abortionfacts.com, viewed March 12, 2017.

30. E. W. Jackson, an African American, Marine Corps veteran, Harvard Law graduate, and practicing preacher from Chesapeake, Virginia, as cited in washingtonpost.com article "Does November matter to the Virginia GOP?" by Mark Plotkin, May 24, 2013.

31. Alcorn, *Pro-Life Answers to Pro-Choice Arguments*, 303.

32. Gregory Curtis speaking of the concentration camp at Dachau in "Why Evil Attracts

Us," *Facing Evil* (Peru, Illinois: Open Court
Publishing Company, 1988), 96.

33. Albert Camus in *The Plague*, (New York:
    Modern Library, 1948), 240, as cited by Karl
    E. Weick in "Small Sins and Large Evils,"
    Facing Evil, 84.

Chapter 9. Help and Support

1. From a Christmas card received December
   2012 from Morning Star Pregnancy Services,
   Harrisburg, Pennsylvania.

2. John C. Willke and Barbara H. Willke, *Why
   Can't We Love Them Both, Questions and
   Answers About Abortion* (Cincinnati: Hayes
   Publishing Company, 1997), 270.

3. Ibid., 271.

4. Ibid.

5. OptionLine, www.optionline.org, www.preg-
   nancycenters.org, viewed March 30, 2017.

6. Ibid.

7. Heidi Murkoff and Sharon Mazel, *What to
   Expect When You're Expecting*, 4th Edition
   (New York: Workman Publishing Company,
   2008), 59.

8. Ibid., p. 60.

9. Ibid., p. 66.
10. Ibid.
11. Ibid.

Chapter 10. Choosing Life

1. Evan Clark, President and Chief Executive Officer of the Department of Commerce Federal Credit Union, April 2008.
2. South Dakota reference to a fetus as part of its informed consent regulations. "Ruling Gives South Dakota Doctors a Script to Read," The Washington Post, July 20, 2008.
3. Francis J. Beckwith, *Defending Life, A Moral and Legal Case Against Abortion Choice* (New York: Cambridge University Press, 2007), 164.
4. OptionLine, www.optionline.org, www.pregnancycenters.org, viewed March 30, 2017.

Ultrasound Imaging

1. An important source of this material is *Prenatal Tests and Ultrasound*, by Elizabeth Crabtree Burton and Richard L. Luciani (Oxford and New York: Oxford University Press, 2012).

Abortion and the Debate on "When Human Life Begins"

1. Kay Elder and Brian Dale, *In-Vitro Fertilization*, Third Edition (Cambridge: Cambridge University Press, 2011), viii.

2. Ibid.

3. Ibid.

4. Randy Alcorn, *Pro-Life Answers to Pro-Choice Arguments* (Colorado Springs: Multnomah Books, 2000), 84.

5. Elder and Dale, *In-Vitro Fertilization*, viii.

6. Alcorn, *Pro-Life Answers*, 88.

7. Elder and Dale, *In-Vitro Fertilization*, viii.

8. Alcorn, *Pro-Life Answers*, 51-56.

9. Ibid., 85.

Prayer for Aborted Children

1. *Prayer for Aborted Children* by Jim Harrison, author of this book.

PHOTO CREDITS
AND SOURCES

1. A Sperm Cell Fertilizing an Ovum. Source: Wikipedia (http://en.wikipedia.org/wiki/ Spermatazoa).

2. Nuclei of the Sperm and Ovum Dynamically Interact to Form a Zygote. Courtesy of the Carnegie Collection (CC No. 8500.1), National Museum of Health and Medicine, Washington, D.C.

3. The Structure of Part of a DNA Double Helix. Source: Wikipedia (http://en.wikipedia.org/ wiki/DNA_double_helix).

4. Eight-Week-Old Embryo. Courtesy of the Carnegie Collection (CC No. 417), National Museum of Health and Medicine, Washington, D.C.

5. 21-Week Fetus Grasping Hand of Surgeon. Courtesy of Michael Clancy. See www. michaelclancy.com.

6. Slave Trade Auction Block. Courtesy of Louisiana State Museum, New Orleans, Louisiana.

7. Japanese American WWII Internment Camp. Courtesy of Topaz Museum, Delta, Utah.

8. Ultrasound of 22-Week Fetus. Source: DHD Multimedia Gallery (Anonymous Contributor).

ABOUT THE AUTHOR

Jim Harrison is a scientist and attorney who lives in Evans, Georgia. He has been married to his wife, Beverly, for more than fifty years. Together they have two wonderful sons, two amazing daughters-in-law, and four grandchildren who are the loves of their lives.

Jim has academic degrees from Montclair State University, the University of Nebraska at Omaha, American University, and the American University Washington College of Law. He has also studied at California State University at Chico and Texas A&M University.

Jim's work experience includes serving as a weather officer in the United States Air Force and as mathematician, physical scientist, meteorologist, and supervisory meteorologist in the National Oceanic and Atmospheric Administration (NOAA). His last position before retirement was Deputy Federal Coordinator for Meteorological Services and

Supporting Research in Silver Spring, Maryland, a responsibility that coordinated operational weather and research activities among fourteen federal agencies.

Since retirement, Jim has attempted to make contributions in the pro-life area concerning unborn children.

CPSIA information can be obtained
at www.ICGtesting.com
Printed in the USA
LVOW11s2235250717
542518LV00001B/59/P